Physical Examination in Cardiovascular Chinese Medicine

Physical Examination in Cardiovascular Chinese Medicine

Dr. Anika Niambi Al-Shura, BSc., MSOM, Ph.D
Continuing Education Instructor
Niambi Wellness
Tampa, FL

Medical Illustrator: Samar Sobhy

ELSEVIER

AMSTERDAM • BOSTON • HEIDELBERG • LONDON
NEW YORK • OXFORD • PARIS • SAN DIEGO
SAN FRANCISCO • SINGAPORE • SYDNEY • TOKYO

Academic Press is an imprint of Elsevier

Academic Press is an imprint of Elsevier
32 Jamestown Road, London NW1 7BY, UK
The Boulevard, Langford Lane, Kidlington, Oxford, OX5 1GB, UK
Radarweg 29, PO Box 211, 1000 AE Amsterdam, The Netherlands
225 Wyman Street, Waltham, MA 02451, USA
525 B Street, Suite 1900, San Diego, CA 92101-4495, USA

Notices
Knowledge and best practice in this field are constantly changing. As new research and experience broaden our understanding, changes in research methods, professional practices, or medical treatment may become necessary.

Practitioners and researchers must always rely on their own experience and knowledge in evaluating and using any information, methods, compounds, or experiments described herein. In using such information or methods they should be mindful of their own safety and the safety of others, including parties for whom they have a professional responsibility.

To the fullest extent of the law, neither the Publisher nor the authors, contributors, or editors, assume any liability for any injury and/or damage to persons or property as a matter of products liability, negligence or otherwise, or from any use or operation of any methods, products, instructions, or ideas contained in the material herein.

British Library Cataloguing-in-Publication Data
A catalogue record for this book is available from the British Library

Library of Congress Cataloging-in-Publication Data
A catalog record for this book is available from the Library of Congress

ISBN: 978-0-12-800120-2

For information on all Academic Press publications
visit our website at **store.elsevier.com**

This book has been manufactured using Print On Demand technology. Each copy is produced to order and is limited to black ink. The online version of this book will show color figures where appropriate.

ELSEVIER Book Aid International **Working together to grow libraries in developing countries**

www.elsevier.com • www.bookaid.org

DEDICATION

The energy and effort behind the research and writing of this textbook is dedicated to my son, Khaleel Shakeer Ryland. May this inspire and guide you through your journey in your medical studies, career, and life.

ACKNOWLEDGMENTS

This is a special acknowledgment to my 7-year medical students at Tianjin Medical University (2012–2013) who served as cardiovascular research assistants. May your future medical careers be successful.

An Qi He
Bin Lin Da
Han Jiang
Chen Hua
Jia Ying Luo
Jun Zhang
Lin Lin
Ming Lu
Nang Zhang
Ping Tang
Hu Si Le
Zhao Tian Man
Wen Xing Ning
Xing Wen Zhao
Tang Ying Mei
Li Ying Ying
Xiong Yong Qin
Ding Yu
Li Yan Jun

CONTENTS

SECTION III EXAM ROOM 1

SECTION IV EXAM ROOM 2

NIAMBI WELLNESS
INTEGRATIVE CARDIOVASCULAR CLINICAL MEDICINE

The companion course for this textbook edition can be found on the Elsevier website and at www.niambiwellness.com.

APPROVING AGENCIES

PROFESSIONAL
NCCAOM
DEVELOPMENT ACTIVITY ®

The course with this textbook is entitled, Integrative Examination in Cardiovascular Disease.

This course is approved by the National Certification Commission for Acupuncture and Oriental Medicine (NCCAOM), and is listed as course #1053-006 for 9 PDA points.

This course is approved by the Florida State Board of Acupuncture, and is listed as course # 20-334888 for 12 CEU credits.

COURSE DESCRIPTION

This course focuses on the basic cardiac patient examination routine using protocols of both Western medicine and TCM.

COURSE OBJECTIVES

- Understand the basics of how to take a cardiac patient history and triage vitals.
- Understand blending of basic Western and TCM methods for examining cardiovascular patients in the clinic.

Getting Started

CHAPTER 1

Introduction

CHAPTER OBJECTIVES

After studying this chapter, you should be able to:

1. List the three important things to collect as an introduction to treating the first-time patient.
2. Describe the four important factors to understand about the patient from the family and present history.
3. Discuss the two reasons mentioned for the female health history.
4. Discuss the two reasons why the patient must report the present medications including any pharmaceutical drugs, Chinese medicines, and nutritional supplements.

1.1 PART 1: THE INTRODUCTION

1.1.1 Name of the Primary or Referring Doctor

In cardiovascular health questionnaires, it is important to focus on specific information for the first step in differentiation in both Western and Chinese medicine. This item is important because of the need to exchange or retrieve health findings for the patient. Information from the referring provider should come before the patient.

1.1.2 Purpose of the Cardiovascular Visit

This item is most important because of the need to focus on the chief complaint.

1.1.3 Pain Ratings

Pain ratings with 0−10 number values.

This item is important because of the need to determine if treatment is indicated, whether the patient should be referred back to the primary medical doctor (MD), emergency medical technician (EMT) assistance, or immediate cardiopulmonary rescuscitation (CPR).

Physical Examination in Cardiovascular Chinese Medicine. DOI: http://dx.doi.org/10.1016/B978-0-12-800120-2.00001-6

1.2 PART 2: THE FAMILY AND PRESENT HISTORY

1.2.1 Lifestyle Factor

Knowing how often these habits are indulged helps to determine the cardiovascular treatment course. In some cases, Chinese medicine may be helpful; however, illicit addictions always complicate or disappoint treatment outcomes.

- Cigarettes
- Alcohol consumption
- Illegal drug use
  - ☑ High blood pressure
  - ☑ Heart attack
  - ☑ Arrhythmia
  - ☑ Diabetes
  - ☑ High cholesterol
  - ☑ Heart failure

1.2.2 Disease History Factor

Family history increases the likelihood of development due to genetics. Incidence in personal history with or without family history can help rule out lifestyle choices that contribute or may be the isolated direct cause.

- ☑ High blood pressure
- ☑ Heart attack
- ☑ Arrhythmia
- ☑ Diabetes
- ☑ High cholesterol
- ☑ Heart failure
- ☑ Stroke

1.2.3 Treatment Procedure History Factor

Patients with a history of these procedures may safely be indicated for TCM treatments.

- ☑ Vascular studies
- ☑ Catheterization
- ☑ High cholesterol
- ☑ Heart surgery

Patients with a history of these procedures may not safely be indicated for TCM treatments.

☒ Pacemaker/defibrillator
☒ Transplantation
☒ Heart surgery

As in all histories, it is necessary to know this information.

• Date of first occurrence/procedure
• Date of recent occurrence/procedure
• Device type: medical implants, pacemaker, diagnostic or monitoring devices

1.3 PART 3: FEMALE HISTORY

Nowadays, heart disease is being diagnosed in women more than in the past. It was once assumed that heart disease in women was rare or nonexistent. Studies are being conducted to identify clear risk factors that set women apart from common ones that are more associated with men.

a. This history can help take a look into how hormonal shifts in life events and aging-related milestones may play a role in the development of heart problems.
   ☑ Year of first menses
   ☑ Number of pregnancies
   ☑ Menopause
   ☑ Hormone replacement (HRT)
b. This history can help take a look into when certain health conditions exclusive to women may have included serious cardiovascular symptoms even if they were temporary.
   ☑ Number of pregnancies
   ☑ High blood pressure during pregnancies or preeclampsia

1.4 PART 4: PRESENT MEDICATIONS

a. This history is important to know because it may give a good idea of the realities of a patient's condition. Also, it helps to avoid prescribing any TCM formulas and supplements which will cause adverse reactions with prescribed drugs.
   ☑ Medicines prescribed for other conditions
   ☑ Current heart medicines

b. This history is important to know, because it may help you understand the stage of recovery for the patient. Also, it helps to document this and share with primary health practitioners who may elect to perform surgical procedures or adjust medicine therapy.
☑ Chinese medicine
☑ Nutritional supplements and alternative substances

1.5 PART 5: CURRENT SYMPTOMS

Make a checklist according to the common symptoms in Western medicine and TCM differentiation.

1.5.1 Cardiovascular Disease and TCM Differentiations

Western medicine	Chinese medicine
Vascular problems	Chest obstruction
Arrhythmia	Blood stasis
Heart failure	Hyperactive yang
Hypertension	Qi or yin deficiency
High cholesterol	Blood stagnation
Coronary heart disease	

Common symptoms:

☑ Chest pain
☑ Palpitations
☑ Leg pain

1.5.2 Respiratory Diseases and TCM Differentiations

Western medicine	Chinese medicine
Cor pulmonale	Phlegm accumulation
Sleep apnea	Qi deficiency
COPD	Yin deficiency
Bronchitis	Wind attack
Pleural effusion	Dampness
Pulmonary infection	
Asthma or emphysema	

Common symptoms:

☑ Snoring
☑ Dyspnea

☑ Wheezing
☑ Prolonged exhalation
☑ Coughing with/without sputum
☑ Chest rattling

1.5.3 Gastrointestinal Diseases and TCM Differentiations

Western medicine	Chinese medicine
Gastroperesis	Qi deficiency
Enteritis	Blood deficiency
Diverticulitis	Yang deficiency
IBS	Damp heat
Bronchitis	
Pleural effusion	
Pulmonary infection	
Asthma or emphysema	

Common symptoms:

☑ Poor appetite
☑ Abdominal pain
☑ Diarrhea
☑ Weight loss
☑ Nausea with/without vomiting
☑ Weight gain
☑ Constipation

1.5.4 Nephrology/Urology: Diseases and TCM Differentiations

Western medicine	Chinese medicine
Nephrotic:	Chest obstruction
Hyperlipidemia	Blood stasis
Acute kidney injury	Hyperactive yang
Cystitis	Qi or yin deficiency
Acute kidney disease	Blood stagnation
Chronic kidney disease	
Nephritic:	
Nephritis	
Renal failure	

Nephrotic or nephritic common symptoms

☑ Edema of lower body
☑ Urinary problems

- ☑ Digestive problems
- ☑ Thirst
- ☑ Short breath
- ☑ Leg cramps
- ☑ Pale skin
- ☑ Dry itchy skin

1.5.5 Neurological Conditions, Events and TCM Differentiations

Western medicine	Chinese medicine
Stroke	Blood deficiency
CVA	Phlegm obstruction
TIA	Blood stasis
Aneurysm	Vascular bi
Hematoma	
Pernicious anemia	

Common symptoms:

- ☑ Headache
- ☑ Seizures
- ☑ Paralysis
- ☑ Tremors
- ☑ Fainting
- ☑ Abnormal gait
- ☑ Memory problems

1.5.6 Blood Diseases, Disorders and TCM Differentiations

Western medicine	Chinese medicine
Anemia	Blood deficiency
Pulmonary embolism	Blood stagnation
Thrombocytopenia	Qi deficiency
Thalassemia	Blood stasis
Endocarditis	

Common symptoms:

- ☑ Light-headed
- ☑ Fainting
- ☑ Bleeding

☑ Short breath
☑ Dry itchy skin
☑ Blood clots
☑ Bruising

1.5.7 Infections and TCM Differentiations

Western medicine	Chinese medicine
Sepsis	Tuo syndrome
Aplastic anemia	Qi and blood deficiency
Endocarditis	Epidemic heat
Viral myocarditis	Wind- cold/ heat dampness

Common symptoms:

☑ Fever
☑ Coldness or chills
☑ Muscle pain
☑ Swollen nodes
☑ Inflammation
☑ Joint pain
☑ Bloody or dark urine

1.5.8 Integumentary Connections to Cardiovascular Diseases and TCM Differentiations

Western medicine	Chinese medicine
Sepsis	Pathogenic heat
Staph infection	Wind syndrome
Endocarditis	Bi syndrome
Viral myocarditis	

Common symptoms:

☑ Rash
☑ Red painful nodes on finger pad
☑ Red painless spots on fingers and toes
☑ Splinter hemorrhages under nails
☑ Localized skin inflammation
☑ Joint pain

NOTES

Log on at www.niambiwellness.com to access the companion course and quiz for Module 1.

The Triage

CHAPTER 2

Weight Measurement

CHAPTER OBJECTIVES

After studying this chapter, you should be able to:

1. Classify lipoprotein disorders in Chinese medicine.
2. Explain the two risk factors linked to heart failure.
3. List the genetic causes of high lipid in some patients.
4. Explain the significance of weight circumference.

In Chinese medicine, lipoprotein disorders are categorized under tanzhuo, xueyu, and xuan yuan and are classified under blood stasis and turbid phlegm obstruction. It may also be mentioned in other categories such as headache, stroke, palpitation, obstruction, and vertigo. It is a pathological deficiency between the liver, spleen, and kidneys, featuring depleted liver yin with hyperactive liver yang (as seen in hypertension).

2.1 PART 1: RISK FACTORS

Certain excess lipids as the cause of weight gain are linked to increased risk for heart failure. Various lab reports may show an imbalance between the HDL being too low and the LDL either being at acceptable levels or too high. Consider whether the cause is due to hypothyroidism, elevated estrogen levels in pre-, peri- or and postmenopausal women, deficient sex hormones in men, and type II diabetes. These conditions tend to be connected with elevated cholesterol levels. In some patients, rare familial genetic mutations may contribute to the poor lipid results.

a. Contributing factors involved in weight gain include:
   - High triglycerides, LDL cholesterol, and total cholesterol
   - Low HDL cholesterol
b. Common risk factors include:
   - Hypertension
   - Coronary artery disease

Physical Examination in Cardiovascular Chinese Medicine. DOI: http://dx.doi.org/10.1016/B978-0-12-800120-2.00002-8

- Stroke
- Metabolic factors that lead to insulin resistance and type 2 diabetes
c. Genetic causes to consider in some patients:
  - Familial hyperchylomicronemia (Type 1)
  - Familial hypercholesterolemia (Type II)
  - Familial hyperlipoproteinemia (Type III)
  - Familial hyperlipoproteinemia (Type IV)
  - Apolipoprotein B (Apo B) isoforms B100 and B48
  - Apolipoprotein E (Apo E) allele mutations E2, E3, and E4

2.2 PART 2: BMI

Body Mass Index (BMI) is a screening method to correlate incidence of obesity.

One reason to use BMI as opposed to weight scales is because the weight of a person alone is not enough to determine risk factor for heart disease. For example, an athlete may have a heavy weight of lean muscle, but the risk factors for heart disease may be quite low. Health workers use several other methods to help clearly determine how much weight plays a role. A common one includes the use of calipers to measure waist skin-fold thickness, and a measuring tape to measure waist circumference. BMI uses a person's height and weight as a measure of body fat to correlate incidence of obesity.

It can be an alternative for direct measures of body fat. The correlation between body fat amount and the BMI number varies according to sex, race, and age. People can use BMI to compare their own weight to a standard and target what is appropriate for themselves.

2.2.1 BMI Table

<19	Underweight
19 – 25	Normal
25 – 30	Overweight
>30	Obese

Things to consider while using a standard BMI for everyone:

- Body fat is necessary and not unhealthy.
- Too much body fat has certain health risks to consider, but everyone is different.
- Females have more body fat than males.
- Older people have more fat than younger people.

*Log on to the companion course for the lesson on BAI vs. BMI

2.3 PART 3: WAIST CIRCUMFERENCE

Medical researchers think BMI does not strongly determine risk factors for high blood pressure, high cholesterol levels, and dangerous blood glucose levels as the measurement of the waist circumference. This belief is held for all patients across sex, race, and age.

2.3.1 Comparing Cardiovascular Risk in Adult Males and Females

Gender	BMI Measurement	Waist Measurement
Adult male	25 overweight >30 obese	89- 101 cm. 35- 40 in.
Adult female	25 overweight >30 obese	83- 94 cm. 33- 37 in.

NOTES

For Chapter 2 tutorial, log on at www.niambiwellness.com to access the companion course for module 2.

CHAPTER 3

Taking the Blood Pressure

CHAPTER OBJECTIVES

After studying this chapter, you should be able to:

1. List the two main tools used to measure and examine the blood pressure.
2. Describe the common and genetic factors to consider in hypertension.
3. Describe the moment during the procedure when the systolic pressure can be measured.
4. Describe the moment during the procedure when the diastolic pressure can be measured.
5. Indicate the blood pressure ranges at each stage of hypertension according to JNC7 classification.
6. Indicate the differences between the average heart rates of athletes, adults, and children of age 12 and above.

Measuring the blood pressure is crucial to detecting and monitoring hypertension. Diet, lifestyle, and psychoemotional factors influence are common factors of development. It may also be necessary to consider whether genes are possible culprits in persistent hypertension. For example, the MYH9 gene is related to nephritic syndrome, and renal hypertension may lead to kidney failure.

- In Bartter syndrome, symptoms present in infants and children.
- In Gitelman syndrome, the symptoms present in children or later in life in adults.
- In Liddle syndrome, hypertension is autosomal dominant and begins in childhood featuring hypokalemia and aldosterone deficiency.
- In pseudohypoaldosteronism type 1, autosomal dominant and recessive forms are first identified in newborn infants and in failure to thrive children.

Physical Examination in Cardiovascular Chinese Medicine. DOI: http://dx.doi.org/10.1016/B978-0-12-800120-2.00003-X

3.1 PART 1: TOOLS

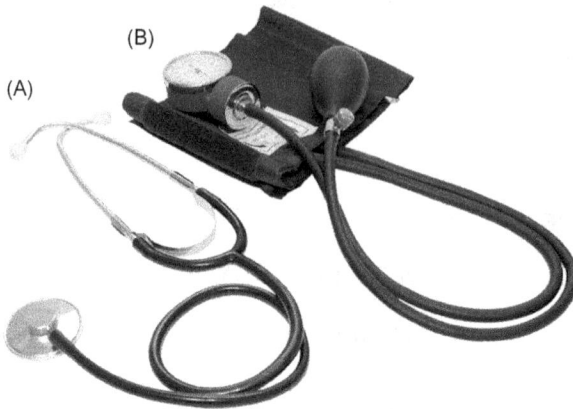

(A) Stethoscope; (B) Sphygmomanometer with cuff. *Copyright © 2014 Anika Niambi Al-Shura. Published by Elsevier Inc. All rights reserved.*

3.2 PART 2: BASIC PROCEDURE

- After the patient has relaxed for about 5 min before taking the blood pressure reading, measure the heart rate (HR). The HR can determine regular patterns.
- Hold the patient's left hand with your right hand. Using the index and middle finger of your left hand, find the radial artery near the wrist. Avoid using your thumb, because your arterial pulsations may interfere.
- Now begin the blood pressure reading. Start the procedure by wrapping the cuff around the left upper arm, while supporting the patient's forearm with your left hand.
- Using the stethoscope, palpate or listen for the brachial pulse located at the anterior medial cubitus fossa and hold it firmly with the thumb of the left hand.
- Use the right hand to operate the pump. Start with the sphygmomanometer gauge reading near 0 mmHg, then moderately pump to inflate the level to above 180 mmHg until the artery has collapsed.
- Then slowly release the air valve to gradually control the expelling of air until the cuff is deflated. Watch as the gauge descends and listen for the first sound which is the systolic pressure. Note where the needle is at this point. The next sounds are blood rushing through the artery as the air continues to be released from the cuff.
- As the beat fades off, it may become louder. As soon as the sound is gone, immediately note the diastolic blood pressure.
- Completely deflate the cuff and remove it. Record the systolic and diastolic pressure.

3.3 PART 3: MEASUREMENT STANDARDS

3.3.1 JNC7 Prevention, Detection, Evaluation, and Treatment of High Blood Pressure Classification

Classification	Systolic	Diastolic
Pre- Hypertension	120–139	80–89
Stage 1 Hypertension	140–159	90–99
Stage 2 Hypertension	>160	>100

3.3.2 Resting Heart Rate

Athletes	40-60
Age 12 to adult+	60-80
Age 12 to adult+ with stages of hypertension	80+

NOTES

For chapter 3 tutorial log on at www.niambiwellness.com to access the companion course for module 2.

CHAPTER 4

Taking the Temperature

CHAPTER OBJECTIVES

After studying this chapter, you should be able to:

1. Describe the purposes of recording body temperature.
2. Explain the problems associated with hyperthermia.
3. Explain the problems associated with hypothermia.

For cardiovascular patients, it is important to keep a clear, simple, and detailed log of body temperature. This record can be included in monitoring and comparing symptoms and episodes which are connected to the changes in season. In addition, temperature changes are often connected with certain viral and bacterial infections, which can be especially dangerous for heart patients.

A good chronicling method of body temperatures should be included on a patient's chart. In some cases, diligent patients who are diagnosed with cardiovascular diseases may want to track origins based on clinical assessments over the years.

4.1 PART 1: PROBLEMS OF HIGH TEMPERATURE

In warmer and hot seasons, heart disease patients are also at risk of heat evil due to hyperthermia. Hyperthermia occurs when the body temperature rises above 100°F and cannot readjust itself to the normal temperature. Symptoms include short breath, angina pectoris, and myocardial infarction. During hot summer months, increased body heat and sweating stresses the cardiovascular system. Some medications can also add to this risk. During this time, patients on cardiovascular medications have an additional challenge. Channel blockers and diuretics can impede the ability of the body to dissipate excess heat. Symptoms from weather changes or internal heat overexposure

Physical Examination in Cardiovascular Chinese Medicine. DOI: http://dx.doi.org/10.1016/B978-0-12-800120-2.00004-1

occur gradually starting with a heat rash, muscle cramping, headache, dizziness, nausea, and vomiting. Later if a patient's temperature rises >105°F, they can suffer heat stroke. Symptoms such as mental confusion, seizures, and possible coma are evidence of heat affecting the neurological system. Acute cardiovascular event symptoms connected with the high heat include short breath, angina pectoris, and myocardial infarction.

4.2 PART 2: PROBLEMS OF LOW TEMPERATURE

Hypothermia can occur during prolonged exposure to cold weather or environments. Some main noticeable symptoms include shivering, mental confusion, sleepiness, slow reactions, and poor coordination. Additional stress is imposed upon the heart to pump faster to circulate blood for warmth.

Health practitioners may notice that patients with symptoms related to coronary heart disease and heart failure will increase clinic visits and emergency room episodes more in the winter than in the summer. Cases of sudden death often occur more in the winter than summer.

4.3 PART 3: PROCEDURE

Core Body Temperature	
The normal body temperature	37°C/98.6°F

4.3.1 Body Reading

A digital thermometer can be used in these places for adults: armpit or mouth. It may seem strange to take an armpit reading for an adult, but in many cases the reading will provide the same results as ear or mouth. The use of a glass or plastic digital thermometer should be a matter of hygienic, monitoring method or protocol preference.

4.3.2 Head Reading

An ear thermometer is instant and effective for detecting body temperature changes. For serious heart patients, an instant ear temperature reading is convenient enough to be taken at different time periods during the day, and more often according to condition:

- Twice during both morning and afternoon for menopausal heart patients. This purpose is also connected with determining hot flashes.
- Twice during night for patients with moderate to severe heart disease. This purpose is connected with increased symptoms during this time of the day.

NOTES

For chapter 4 tutorial log on at www.niambiwellness.com to access the companion course for module 2.

Module Review Questions

1. What are the common cardiovascular risk factors which are involved with excess weight?
2. List the genetic disorders associated with excess lipids and hypertension.
3. What is the significance of BMI measurements?
4. What is the significance of measuring waist circumference?
5. How is the MYH9 gene connected to hypertension?
6. What genetic diseases in children may predispose them to hypertension?
7. Where should the stethoscope be placed to listen to the sounds while taking the blood pressure?
8. When is recording body temperature most significant?
9. When does hyperthermia usually occur?
10. When does hypothermia usually occur?

Log on at www.niambiwellness.com to access the companion course and quiz for Module 2.

Exam Room 1

CHAPTER 5

Physical Exam Basics

CHAPTER OBJECTIVES

After studying this chapter, you should be able to:

1. Explain three things to be noticed during patient observation.
2. List the three patient positions during examination.
3. Explain the comparison between pulse qualities in Chinese and Western medicine.
4. Explain steps in cardiac palpation.
5. List the heart sounds best heard with the diaphragm of the stethoscope.
6. List the heart sounds best heard with the bell of the stethoscope.
7. Explain steps in respiratory palpation.

The physical examination is the crucial section of the clinical visit. The health practitioner must determine whether the patient is indicated for Chinese medicine treatments alone, integrated with Western medicine, or Western medicine alone.

There are limitations on which cardiovascular disease conditions are indicated for Chinese medicine treatments. Scope of practice is determined by local laws in the United States and in other countries. In the United States, if a patient presents with AHA/ACC stage D and NYHA stage 4 heart disease levels, then the patient may require hospitalization. A patient with an advanced condition which frequently needs immediate or emergency assistance is usually not indicated for Chinese medicine treatments. The reason is that the likelihood of treatment effectiveness diminishes considerably with poor patient condition and prognosis. An existing patient who does not respond to Chinese medicine treatments alone may need to integrate with or transition solely to Western medicine treatments and drugs. Referral to Western medicine

Physical Examination in Cardiovascular Chinese Medicine. DOI: http://dx.doi.org/10.1016/B978-0-12-800120-2.00005-3

treatment may prolong and improve the life of the patient. In some cases, the recovering patient can be transitioned in the reverse order back to Chinese medicine for prevention monitoring.

5.1 PART 1: OBSERVATION

Evaluating a patient with heart problems should start with a movement observation. This identifies related problems throughout the body. First of all, a lot of information about a patient can be gained from watching a patient walk into a room. Next watching them stand, walk, and climb upon or off of the exam table. It is especially necessary to do this for elderly patients. If the patient is able, ask them to walk into the exam room and stand, while noticing the gait. Next have them mount the exam table. Stand in a position to assist if they have trouble. Note the following observations:

5.1.1 Balance
The standing or walking patient may kind of veer off to one side or lack a sense of equilibrium in pace, by shuffling, or taking short steps. Climbing upon the table may result in falling off. This is characteristic in stroke, transient ischemic attack (TIA), hypertension, old cerebrovascular accidents, and diabetes types.

5.1.2 Walking Gait
The walking patient may start off slow and then accelerate and finally lose control of balance. Also some patients move while compensating for joint pain or range of motion limitations. This is characteristic of degenerative joint disease, inflammation, and neurological diseases.

5.1.3 Arm and Leg Movement
Patient may not move freely due to contracture.

- This is characteristic in stroke, inflammation, and other neurological diseases.
- Heel to toe test is used only for neurological evaluation.

5.2 PART 2: PREPARATION

5.2.1 Draping and Positioning

Physical examination of a heart patient should include observation, palpation, and auscultation. Percussion is used sometimes to inspect the abdomen but not used in the cardiac exam. Help the patient upon the exam table into a common position during the cardiovascular exam. The positions include: dorsal recumbent, Fowler's, supine, or Trendelenburg's. For pulses, the Fowler's position (face up) is best. The patient position on the table should be supine with the upper body elevated 30°–45°. The patient should be completely naked under a robe or gown, which is open in the front. Female patients should remove the bra. The patient draping is important because it provides modesty. All areas should be covered except for the area to be examined. The room should be quiet with plenty of light for listening and observing. Always explain what you are doing in the exam. Enlist a female patient's assistance by asking her to lift her covered breast during the precordial palpation and at auscultation.

5.2.2 Pulse Inspection

Pulse Inspection	
Carotid pulse	Under the angle of the jaw
Brachial pulse	At the anterior- medial cubitus fossa of the outstretched arm
Radial pulse	At the wrist
Abdominal pulse	Ipsilateral lower abdomen
Femoral pulse	Beneath the inguinal ligament medial to ASIS
Popliteal pulse	Posterior to medial malleolus
Posterior tibial	along the edge of the medial malleolus
Dorsalis pedis	lateral to the extensor hallucis tendon (top of foot)

5.2.3 Carotid Artery Palpation

This exam detects atherosclerosis, severe impairment of the ventricles, and aortic stenosis.

To locate the carotid artery, gently slide your fingers along the side of the trachea at the level of the thyroid cartilage. Palpate one side at a time.

5.2.4 Jugular Vein

When assessing the right jugular venous pressure (JVP), you view the internal jugular vein by asking the patient to turn the head slightly. You should notice the pulsation of the vein beneath the belly of the sternocleidomastoid muscle.

If gentle pressure on the right costal area of the abdomen produces a JVP elevation of more than 5 s, the patient may have hypertension, right ventricular failure, valvular stenosis, or regurgitation. If you cannot see the vein you must do the hepatojugular reflex maneuver. To do this, slightly push the side of the head as the patient resists. A raised vein may be a sign of right heart failure.

5.2.5 Popliteal Pulse

The popliteal pulse of a single knee should be palpated using all fingers of both hands except the thumbs, which are for bracing the sides of the knee. The fingers fit on the medial borders of the fibula and tibia. Pulse is felt when knee is slightly bent.

5.2.6 Dorsalis Pedis Pulse

This pulse is located on top of the foot extending between the big and middle toes. It can be palpated while the foot is extended.

5.2.7 Pulse Qualities

These qualities connect Chinese and Western medicine together during the examination.

Quality	Chinese medicine symptoms	Western medicine symptoms
Rapid pulse	heat, yang rising or yin deficiency	tachycardia, infection, hyperthyroidism, anxiety, infection, anemia, or arterio-venous fistula.
Irregular pulse	heart qi deficiency, qi stagnation	premature beats, and a completely irregular pulse implies the presence of atrial fibrillation.
Slow pulse	pathogenic cold, yang deficiency	bradycardia, hypothyroidism, heart block, propranolol or other heart drug effects.
Absent pulse	blood vessel bi	impaired blood flow, sepsis, autoimmune disease, rheumatic disease, Takayasu Arteritis

5.2.8 Chest Inspection

Notice if the chest is flat or concave as in pectus excavatum. These are characteristic features in emphysema. Notice if the chest is barrel shaped with trapped air in the lungs. This is a characteristic in emphysema, COPD, and other respiratory problems. Obesity and respiratory problems which enlarge the circumference of the chest may require extra effort during examination.

5.2.9 Abdominal Examination

This method is important for serious patients with advanced cardiac symptoms. Inspection detects whether there is tissue or organ pathology. Rebound tenderness near the lower right quadrant may indicate Rovsing's sign or appendicitis, other infections, or female reproductive organ problems. Percussion over the stomach should produce a hollow tone. Percussion over the rest of the abdomen should produce a dull sound. Percussion over the liver with problems produces pain. Percussion over a bloated abdomen produces tympany and may indicate gas bloating. Palpating a doughy or fluid quality of the tissue under the skin may indicate kidney problems. Gently rocking or tapping the abdomen may also produce waves indicating presence of fluid.

5.3 PART 3: CARDIAC PALPATION

5.3.1 Precordial Examination

This exam detects palpable cardiac problems. While the patient is on the exam table lying on the back, help them to turn on their side into the left lateral decubitus position. This position brings the heart closer to the chest wall and makes the point of maximum impulse (PMI) more obvious. Stand behind the patient and cover the chest area of the heart with the palm of your right hand. It should be positioned at about the mid-clavicular line at the fifth intercostal space where the PMI is located.

Abnormal sensations:

- Pulse transmission
- Thrills

Pulse transmission is a vigorous sensation which is associated with mitral regurgitation due to ventricular hypertrophy.

A thrill is a vibration due to the turbulent flow through a valve, suggesting a valve abnormality.

5.4 PART 4: AUSCULTATION

View chapter 5 in the course. List the important areas of auscultation.

1. _____
2. _____
3. _____
4. _____
5. _____

5.4.1 Normal Sounds

Notice sounds produced during systole and diastole. These sounds occur because of the mitral and tricuspid valves, aortic and pulmonic valves.

5.4.2 S2 Is Loudest

First, place the stethoscope diaphragm firmly over the second right intercostal space.

View chapter 5 in the course. List the important area of auscultation.

6. _____

Next slide it medially to the other side and listen where?

View chapter 5 in the course. List the important area of auscultation.

7. _____

5.4.3 Splitting S2 Is Loudest

Move the stethoscope down to which two areas?

View chapter 5 in the course. List the important area of auscultation.

8. _____
9. _____

5.4.4 S1 Is Loudest

Slide down the sternum to the right fourth intercostal space, then to the left fifth intercostal space to listen which section?

Finally, slide the diaphragm to which section? located at the left fifth intercostal space, at the mid-clavicular line?

View chapter 5 in the course. List the important area of auscultation.

10. _____

11. _____

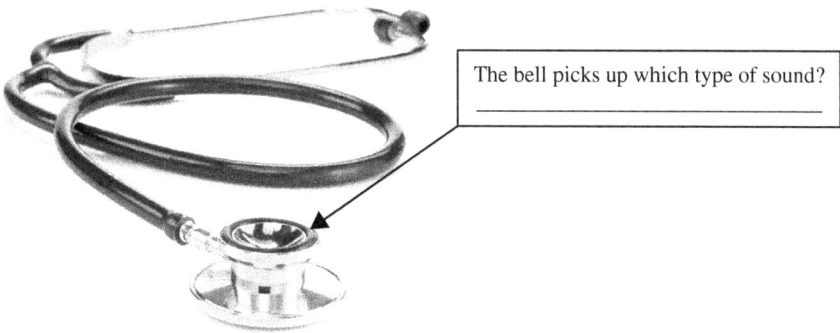

The bell picks up which type of sound?

_____

5.4.5 Summation Gallops

These are S3 and S4 sounds which are also known as extra heart sounds. They are thought to be normal up to age 30. However, in older patients, this represents a pathologic condition. These sounds are soft, so you must roll the patient to the left lateral decubitus position and ask them to exhale. The heart will be closer to the chest wall which increases the sound.

5.4.6 S3 Is Loudest

An S3 sound is associated with left ventricular failure.

The sound is created by an early diastolic filling, when the atrial blood jams into an overfilled ventricle.

View chapter 5 in the course. List the important area of auscultation.

12. _____

5.4.7 S4 Is Loudest
View chapter 5 in the course. List the important area of auscultation.

13. _____

The S4 sound is associated with left ventricular hypertrophy from prolonged hypertension. The sound is created by atrial blood entering a noncompliant left ventricle.

5.4.8 Murmurs
In adults, murmurs are sounds as a result of turbulent blood flow across damaged or prolapsed valves. They can occur during systole or diastole. The sound quality is noted using the Levine Scale with an intensity level between 1 and 6.

5.4.9 Levine Scale

1	Murmur is only audible with careful listening.
2	Murmur is immediately audible but faint.
3	Murmur is clearly audible without palpable thrill.
4	Murmur is clearly audible with a palpable thrill.
5	Murmur is clearly audible with a palpable thrill, and can be heard with light stethoscope touching.
6	Murmur is clearly audible with a palpable thrill, and can be heard without stethoscope touching.

5.4.10 Regurgitation
Blood leaking backward due to prolapsing and therefore insufficiently closing leaflets.

View chapter 5 in the course. List the important area of auscultation.

14. _____

- Mitral valve prolapse
- Aortic regurgitation

Tricuspid regurgitation is associated with elevated pressure transmitted to the right side of the heart.

5.4.11 Stenosis
Blood pushing across stiff valve leaflets that fail to open properly.

- Aortic stenosis
- Mitral stenosis

5.4.12 Carotid Bruit
After chest auscultation, return to the carotid artery to listen for bruit.

5.4.13 Bruits
Listening for this can be done at any artery to detect the presence of thrombosis, occlusion, or inflammation. Bruits may be associated with the presence or recovery from atherosclerosis. The sound of a bruits is the turbulent flow through the artery which is at least 50% occluded.

As the artery becomes more narrow, the sound increases until the narrowing becomes more serious. Then the sound begins to reduce. The sound of bruits is the turbulent flow through the artery which is at least 50% occluded.

To listen for bruit, have the patient briefly hold their breath, then place the diaphragm gently over each carotid area. A bruit gives a "woosh" sound.

5.4.14 Valsalva Maneuver
This helps detect heart abnormalities, treat abnormal heart rhythms, or relieve chest pain.

The seated patient squeezes the nose and mouth shut, while blowing out and bearing down as if having a bowel movement. This raises the heart rate and blood pressure.

Auscultation or echocardiography during the Valsalva maneuver can detect any changes in rate, sound, and pressure, which brings out any latent heart diseases.

The Valsalva maneuver also corrects atrial tachycardia, forcing the heart to beat more slowly, or diminishes chest pain in mild coronary artery disease.

5.5 PART 5: RESPIRATORY PALPATION

Physical examination of the lungs should include observation, palpation, and auscultation.

Percussion is used in the respiratory exam.

5.5.1 Observation

The respiratory exam should be performed while the patient is seated on the table.

First, help the patient sit up and swing both legs to position them hanging off the side of the table. A lot of information about a patient can be gained from watching a patient breathe face to face.

- Notice if the breathing seems shallow, deep, regular, or irregular.
- Notice if the breathing appears to be distressed or labored. The use of the accessory muscles is a sign of respiratory difficulty.
- Notice the color of the lips and nail beds for pale or blue color, which is a sign of cyanosis. This means insufficient amount of oxygen.
- Notice if the patient breathes through pursed lips, whispers or speaks softly or with a rasp.
- Notice if there is any wheezing, gurgling, fremitus, or crackles in the breathing. These are heard in pleural effusions and consolidation.
- Next, stand behind the patient, partially to the side.
- Then move the draping away to expose the back.

5.5.2 Palpation

Touching the back is good for identifying areas of pain and skin temperature changes.

Notice if the spine is curved with kyphosis along with a concave chest. These are characteristic in emphysema and COPD.

5.5.3 Percussion

This method detects whether the lungs are clear, filled with air or fluid. Place last two phalanges of your left middle finger on the patient's back. Tap the joint twice with the middle and index fingers of the right hand to produce sound. Slide around to repeat this method in other areas of the lung field, to notice changes in tone quality. Avoid the scapula and spine. Notice sound tones similar to thumping a melon, drum, or solid.

5.5.4 Tone Qualities
- Normal lungs produce a tone like thumping a melon.
- Lungs with chronic or acute diseases that trap air produce tympany, a tone like a drum.
- Lungs filled with fluid, pleural effusion, and pneumonia produce a tone like tapping a solid surface.

5.6 PART 6: AUSCULTATION

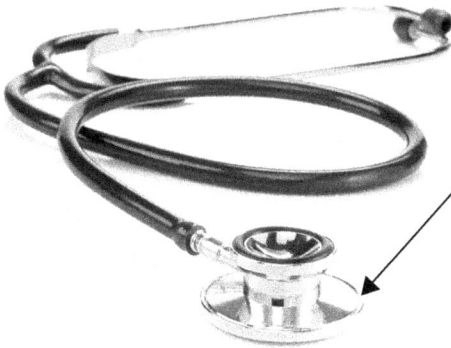

Which type of sound does the diaphragm pick up?
_____

Have the patient breathe through the mouth and exaggerate the slow and deep breaths as you listen around the lung field. This helps to calm the patient and recruit their cooperation. It also helps to identify quality of the lung sounds.

5.6.1 Normal Sounds
This is known as tracheal and vesicular sounds. There is a soft unobstructed sound as air moves from the alveoli through the trachea.

5.6.2 Abnormal Lung Sounds
5.6.2.1 Wheezing
This sound is a kind of whistling noise as a result of air being forced through which area of the respiratory system? Visit the course to view the answer.

1. _____

5.6.2.2 Stridor

This sound is a kind of airy, choppy musical whistle and is caused by an obstruction. There may be a tumor, bolus, or another foreign object and is usually considered a medical emergency.

Where does this sound come from? View the course for the answer.

2. _____

5.6.2.3 Crackles

This sound is also known as rales and is caused by fluid accumulation. Where does this sound come from? View the course for the answer.

3. _____

This sound is often heard in pneumonia, pleural effusion, and congestive heart failure.

5.7 PART 6: AUSCULTATION

Supplement: Cardiac and breathing sounds can be heard on the companion websites.

NOTES

For chapter 5 tutorial log on at www.niambiwellness.com to access the companion course for module 3.

Module Review Questions

1. What is the clinical significance of a patient's ability to maintain balance?
2. What pulses are palpated during the preparation?
3. What two abnormal sensations are detected during the precordial exam?
4. Discuss the normal sounds heard using the bell of the stethoscope.
5. Discuss the abnormal sounds heard using the bell of the stethoscope.
6. Discuss the normal sounds heard using the diaphragm of the stethoscope.
7. Discuss the abnormal sounds heard using the diaphragm of the stethoscope.

Log on at www.niambiwellness.com to access the companion course and quiz for Module 3.

Exam Room 2

CHAPTER 6

Tongue and Pulse Diagnosis

CHAPTER OBJECTIVES

After studying this chapter, you should be able to:

1. Explain the significance of the tongue and pulse diagnosis.
2. Compare tongue quality with pulse quality.
3. Compare related diagnoses of Chinese and Western medicine according to pulse quality.

6.1 CLINICAL ASSESSMENT

Tongue and pulse quality are important diagnostic tools in Chinese and Western medicine. However, in cardiovascular Chinese medicine, tongue diagnosis is not an important part of the patient examination because of the nature of diseases. Pulse quality becomes more important because it will indicate changes in the condition of the patient and progression of the disease or treatment. Once a patient has been diagnosed based on patterns of fixed signs and symptoms and other diagnostic testing results, the treatment principle and treatments become more standardized.

The following images are more of a representation that the tongue can appear mildly or severely affected, no matter whether the condition is serious or not.

6.2 CASE #1

6.2.1 Tongue Quality

1. _____
2. _____
3. _____
4. _____

View the answers in the course.

Physical Examination in Cardiovascular Chinese Medicine. DOI: http://dx.doi.org/10.1016/B978-0-12-800120-2.00006-5

Assessment	Pulse Quality	Related Diagnoses
Chinese Medicine	Weak Slow Uneven Deep	Chest obstruction Heart qi deficiency Yang deficiency
Western Medicine	Irregular	Coronary heart disease Congestive heart failure

6.3 CASE #2

6.3.1 Tongue Quality

1. _____
2. _____
3. _____
4. _____

View the answers in the course.

Assessment	Pulse Quality	Related Diagnoses
Chinese Medicine	Weak Slow Uneven Deep String Forceful	Chest obstruction Heart qi deficiency Yang deficiency Blood stagnation
Western Medicine	Irregular Pulsus alterans	Coronary heart disease Congestive heart failure Hypertension

6.4 CASE #3

6.4.1 Tongue Quality

1. _____
2. _____
3. _____
4. _____

View the answers in the course.

Assessment	Pulse Quality	Related Diagnoses
Chinese Medicine	Wiry Thready Rapid	Heat retention due to yin deficiency Epidemic heart heat Deficiency of qi and yin
Western Medicine	No significance	Viral myocarditis Congestive heart failure

6.5 CASE #4

6.5.1 Tongue Quality

1. _____

2. _____

3. _____

4. _____

View the answers in the course.

Assessment	Pulse Quality	Related Diagnoses
Chinese Medicine	Weak Slow Uneven Deep Thready Forceful	Ren and chong maladjustment Qi deficiency Blood deficiency
Western Medicine	Irregular Pulsus parvis	Hypertension Congestive heart failure

NOTES

For chapter 6 tutorial log on at www.niambiwellness.com to access the companion course for module 4.

CHAPTER 7

Putting It All Together

7.1 PART 1: CASE STUDY #1

The 41-year-old male cardiac patient presented with chest pain that radiates to the shoulders and back, cold sensation in the limbs, the chest was stuffy, and the breath was short.

Patient was diagnosed a few years ago, which is not currently being managed.

1. What is the Western medicine diagnosis?

   _____

2. What is the TCM treatment principle?

   _____

View the course for the answers.

Personal History	
Family History	heart failure
Female History	N/A
Medications	multivitamins, aspirin

Physical Examination	
Weight	BMI 35
Blood Pressure	150/110
Heart Rate	100
Temperature	100°F
Tongue	The tongue was dark
Pulse	Tight (TCM)

Physical Examination in Cardiovascular Chinese Medicine. DOI: http://dx.doi.org/10.1016/B978-0-12-800120-2.00007-7

Auscultation	
Heart	S4 sound
	S3 sound
	Carotid bruits
Lungs	Normal

7.2 PART 2: CASE STUDY #2

The 42-year-old female cardiac patient presented with precordial pain, palpitations, lassitude, spontaneous perspiration, shortness of breath, and dry mouth.

Patient has no previous diagnosis with heart disease but may be?

1. What is the Western medicine diagnosis?

   _____

2. What is the TCM treatment principle?

   _____

View the course for the answers.

Personal History	
Family History	death of a male grandparent
Female History	irregular but heavy menses
Medications	multivitamins, nutritional supplements, herbs and birth control pills

Physical Examination	
Weight	BMI 25
Blood Pressure	115/70
Heart Rate	67
Temperature	95°F
Tongue	pale and dry
Pulse	weak and tight (TCM); positive JVP

Auscultation	
Heart	Normal
Lungs	Normal

7.3 PART 3: CASE STUDY #3

The 50-year-old male cardiac patient presented with palpitations, anger, exhaustion, history of nocturnal emission, dry stools, and red tongue.

Valsalva maneuver relieved symptoms.

1. What is the Western medicine diagnosis?

   _____

2. What is the TCM treatment principle?

   _____

View the course for the answers.

Personal History	
Family History	father has heart disease
Female History	NA
Medications	western heart drugs

Physical Examination	
Weight	BMI 30
Blood Pressure	150/100
Heart Rate	107
Temperature	100.5°F
Tongue	pale and swollen
Pulse	rapid (TCM)

Auscultation	
Heart	S3 S4
Lungs	Normal

7.4 PART 4: CASE STUDY #4

The 39-year-old female cardiac patient presented with headache before and after menses, headache, dizziness, sternal–costal distension, restlessness, insomnia, and fluctuations in blood pressure.

1. What is the Western medicine diagnosis?

_____

2. What is the TCM treatment principle?

_____

View the course for the answers.

Personal History	
Family History	mother diagnosed with heart disease
Female History	regular menses
Medications	western medicine for headache

Physical Examination	
Weight	BMI 20
Blood Pressure	135/85
Heart Rate	97
Temperature	97°F
Tongue	pale
Pulse	tight and weak with connection between the three pulse positions

Auscultation	
Heart	Normal
Lungs	Normal

7.5 PART 5: CASE STUDY #5

The 31-year-old male cardiac patient presented with oily skin, dizziness, heaviness, coughing up of phlegm, and a bitter mouth taste.

1. What is the Western medicine diagnosis?

_____

2. What is the TCM treatment principle?

_____

View the course for the answers.

Personal History	
Family History	no family history
Female History	NA
Medications	western medication for hypertension

Physical Examination	
Weight	BMI 22
Blood Pressure	145/97
Heart Rate	85
Temperature	96°F
Tongue	greasy coat
Pulse	smooth and rolling

Auscultation	
Heart	Normal
Lungs	Normal

7.6 PART 6: CASE STUDY #6

The 75-year-old male patient presented with dyspnea, a dull expression, purple lips, lethargy, pain in the chest and limbs, and phlegm heard rattling around in the chest.

1. What is the Western medicine diagnosis?

   _____

2. What is the TCM treatment principle?

   _____

View the course for the answers.

Personal History	
Family History	no known family history of heart disease
Female History	NA
Medications	western medicine heart medicines

Physical Examination	
Weight	26
Blood Pressure	160/111
Heart Rate	115
Temperature	99°F
Tongue	bright red
Pulse	rolling and wiry

Auscultation	
Heart	Gallops
Lungs	Crackles Wheezing

For chapter 7 tutorial log on at www.niambiwellness.com to access the companion course for module 4.

Module Review Questions

1. List the three important things to collect as an introduction to treating the first time patient.
2. Describe the four important factors to understand about the patient from the family and present history.
3. Discuss the two reasons mentioned for the female health history.
4. Discuss the two reasons why the patient must report the present medications including any pharmaceutical drugs, Chinese medicines, and nutritional supplements.
5. In Chinese medicine, what are lipoprotein disorders often classified under?
6. Explain the two risk factors linked to heart failure.
7. List the genetic causes of high lipid in some patients.
8. Explain the significance of weight circumference.
9. List the two main tools used to measure and examine the blood pressure.
10. What are the common and genetic factors to consider in hypertension.
11. Describe the moment during the procedure when the systolic pressure can be measured.
12. Describe the moment during the procedure when the diastolic pressure can be measured.
13. Indicate the blood pressure ranges at each stage of hypertension according to JNC7 classification.
14. Indicate the differences between the average heart rates of athletes, adults, and children of age 12 and above.
15. Describe the purposes of recording body temperature.
16. Explain the problems associated with hyperthermia.
17. Explain the problems associated with hypothermia.
18. Explain three things to notice during patient observation.
19. List the three patient positions during examination.
20. Explain the comparison between pulse qualities in Chinese and Western medicine.
21. Explain steps in cardiac palpation.
22. List the heart sounds best heard with the diaphragm of the stethoscope.
23. List the heart sounds best heard with the bell of the stethoscope.
24. Explain steps in respiratory palpation.
25. Explain the significance of the tongue and pulse diagnosis.
26. Compare tongue quality with pulse quality.
27. Compare related diagnoses of Chinese and Western medicine according to pulse quality.

Log on at www.niambiwellness.com to access the companion course and quiz for Module 3.

This also concludes the Integrative Anatomy and Pathophysiology in TCM Cardiology course. It is strongly suggested that you log onto the courses at the companion websites and review the course modules. Next, submit course documents and complete the final exam.

Upon passing the exam, you will receive completion certificates which include your name and practice license number, along with the specific number of credit hours awarded for this course. Electronic transmission of CEU and PDA credits will be sent to NCCAOM and your state medical board.